ELEMENTRY INTERNAL COMBUSTION ENGINE

BY

AFOFU OLUWAFEMI SULAIMON
HND (MECH IB POLY 2014) MIANG, MIRED.MNIME

PREFACE

This book is intended for student of engineering degree and diploma courses and scope in sufficiently wide range, to cover the variation in syllabus and standards of various universities and polytechnics and technical colleges. It will effectively serve as hand book, supplementing text of library reference.

AFOFU O. SULAIMON

DEDICATION

This book is dedicated to Almighty Allah, and also to the technological advancement of mankind.

ABRRREVATIONS

Generally in this hand book abbreviations will be used frequently, student will find this list useful for quick reference.

- ICE   internal combustion engine
- SIE   spark ignition engine
- CIE   compression ignition engine
- TDC   top dead center
- BDC   bottom dead center
- SAE   society of automobile engineers
- BORE   diameter of engine cylinder
- IO   Inlet valve open
- IC   Inlet valve close
- EO   Exhaust valve open
- EC   Exhaust valve close

TABLE OF CONTENT

1.0 INTRODUCTION

Automobile engine converts energy contained in fuel into relatively efficient and inexpensive transportation. Most of the modern automobile is design with internal combustion engine which are used in cars, trucks, and buses. The name internal combustion (I.C) engine is mainly classified into two which include the spark ignition engine (SIE) and the compression ignition engine (CIE). The SIE works with the mixture of air and fuel which is ignited by a spark plug and CIE works with compressed fuel at high temperature. The I.C engine is divided into two and four stroke engines. In four stroke engines the piston accomplishes four distinct strokes for every two revolution of the crankshaft. In two strokes engine there are two distinct stoke in one revolution. Variation of this engine types are used in today's automobile and include differences in number of cylinders, size and cylinder arrangement.

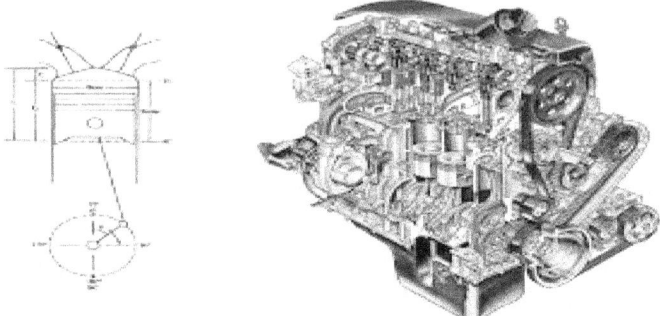

Figure 1.0 Internal combustion engine

1.1 THE EVENT ABOVE THE PISTON

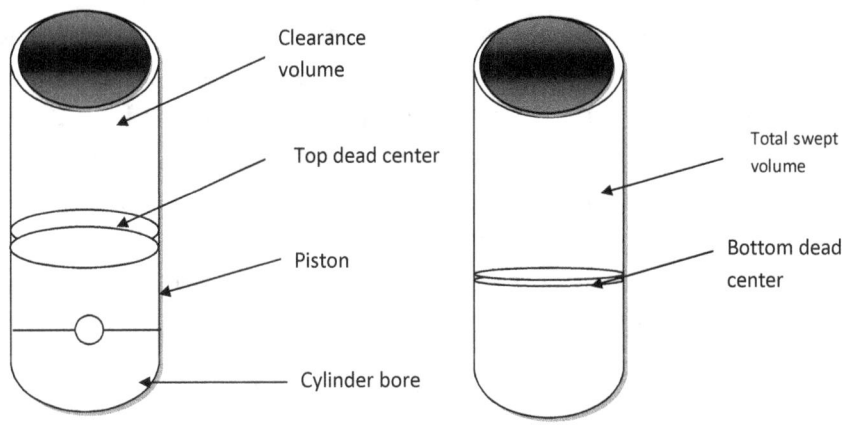

Figure 1.1 Events above the piston

1.2 ENGINE CONSTRUCTION

IC engine consist of a few large components and many small ones, most engines have more than one cylinder, It depends on the capacity of the engine produced by the manufacture. These are the major components that form up an engine.

1. Cylinder block
2. Cylinder head
3. Crankshaft
4. Piston
5. Connecting rod
6. Fly wheel
7. Cam shaft
8. Bottom tray/sump

1.2.1 CYLINDER BLOCK

This is a major component which the engine is built upon and the bore is built within were the piston is been housed. The cylinder block is mostly constructed with cast iron and aluminum alloy and variation depend on the manufacturer. The cylinder block must be rigid to hold the bore relatively to each other and hold the crankshaft in place. The crankshaft runs at right angle to the bore and being retained in the main bearing. The block form one semi circular half of the bearing and semi circular cap from the other half. The cylinder block has the engine mountain attached to outside of the motor vehicle.

Figure 1.1   Cylinder block

1.2.2 CYLINDER HEAD

The cylinder head as the name implies it's on top of the cylinder block and it cover the cylinder block to for a combustion chamber with the machined top of the piston. The cylinder is shaped so that the combustion chamber is actually in the cylinder head. The cylinder head houses the valve, valve spring and the rocker arm, this part of the valve works via a push rods.

Sometimes the camshaft is fitted directly into the cylinder head and operates on the valve without rocker arm. This arrangement is called an overhead camshaft arrangement. The cylinder head can also be manufactured with cast iron or aluminum alloy due to the high temperature that occurs within the region. The cylinder head is been held to the cylinder block with high tensile steel stud. The joint between the cylinder block and the cylinder head must be air gas tight so that there will be complete combustion. This is achieved by using a cylinder to gasket in between the cylinder block and cylinder head.

Figure 1.2 cylinder head

1.2.3 CRANKSHAFT

the crankshaft plays an important role in IC engine it work in conjuction with the connecting rod, converts the reciprocating motion of the piston to the rotary motion needed produce a certain capacity by the engine. It made from carbon steel which is alloyed with a small propotion of nickel. The main bearing jounals is fitted into the cylinder block and the big end jounals align with the connecting rods. The rear end of the crankshaft is attached to the flywheel and at the front end are the driving wheel for the timing gears, water pumps, and the cooling fan.

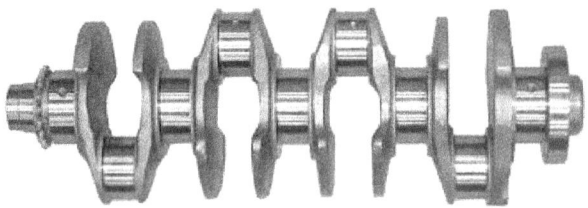

Figure 1.3 Crankshaft

1.2.4 PISTON

The piston reciprocates in the cylinder bore going form top dead center (TDC) to bottom dead center (BDC). Its purpose it to keep gases in the cylinder bore tightly sealed in place and to transmit the pressure produced when gases are burnt during the power stroke to the gudgeon pin. Piston are made from aluminum alloy, which is light, strong and good conductor of heat. They run at a speed up to 13m/s with a temperature range being as high as 2000^0c at the crown and as low as freezing point where the gudgeon pin fits. Piston sizes vary due to the capacity and the size of the engine and also work to complete the gas tighten with **the ring**.

Figure 1.4 piston

The piston is mainly classified into four types.

1. PLAIN, SOLID, OPEN-ENDED

This types of very strong but heavier and has its skirt connected to it head region all the way round. It is necessary to allow clearance especially at the top of the skirt to avoid seizure.

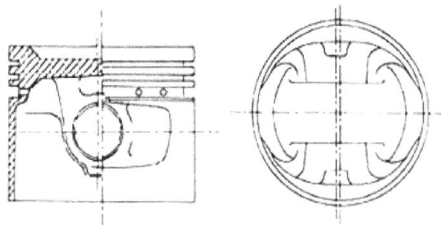

Figure 1.5 plain, solid, open-ended

2. TRANVERSE SLOT,OPEN- ENDED

Such a piston is thermally compensating and can use usually be operated at smaller clearance at the top of the skirt in the interest of noise reduction.

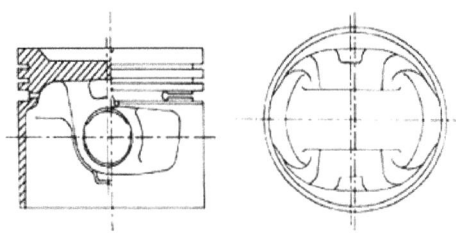

Figure 1.6 transverse slots, open ended

3. SOLID SKIRT SLIPPER

This type of piston has no thermal slot but the skirt is heavily cut away in the pin axis to reduce weight and avoid swaying, it is also rigid and usually needs greater clearance than other types, but this type is the strongest.

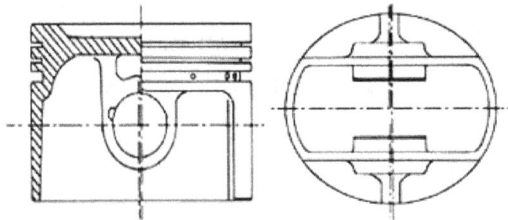

Figure 1.7 solid skirt slipper

4. TRANVERSE SLOT SLIPPER

Contrary to the solid skirt slipper type, this has a cast of milled transverses slot in the oil ring grove for thermal compensation. Smaller clearance at the top of the skirt is required, in the interest of less noise.

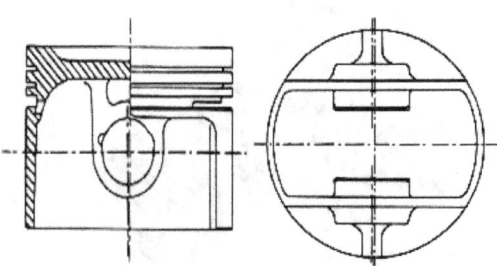

Figure 1.8 Tranverse slot slipper

1.2.5 CONNECTING ROD

The major function of the connecting rod is to connect the piston to the crankshaft, together with the crankshaft they form a simple mecanism that converts reciprocating motion into rotation motion, the might also convert in opposite side it help the piston to achive the up and down stroke in the bore. It is usually made with cast iron , a bush made from a soft metal, such as bronze is use for this joint. The smaller part go to the piston and the other end to the crankshaft.

1.7 FLYWHEEL

Flywheel is a large- diameter, heavy disk, usually constructed with cast iron. It is bolted to the engine crankshaft. The flywheel smoothies out, damps, engine vibration caused by firing pluses. It also acts as a friction surface and heat sink for one side of the clutch disk. The teeth around the circumference of the flywheel form a ring gear, which when engaged to the starter motor pinion gear are used to start the engine.

Figure 1.9 fly wheel

1.2.6 VALVE TIMING

The crankshaft and the camshaft rotations also operates the valve of the engine , the opening and closing of the valves both the inlet and the exhaust are well timed on the cam shaft rotations. the valve mechanism operates on the rotation of the camshaft lobes that open and closes the valve. The engine will and efficient performance when the appropriate mass of air and fuel enters into the combustion chamber and burn effectively. The opening and the closing of the piston in relation to the piston and crankshaft position is called valve timing.

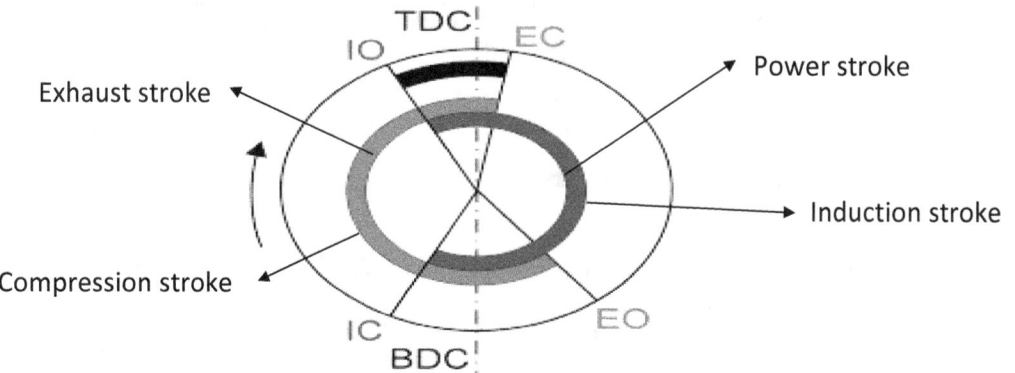

Figure 1.9.1 valve timing

1.2.7 ENGINE FIRING ORDER

The order at which the spark occurs in each cylinder is called the firing order. In order to obtain the best balance, the four the cylinder in-line engine employed crankshaft arrangement where the front and rear piston are in TDC and BDC while the center pistons are in BDC as shown below.

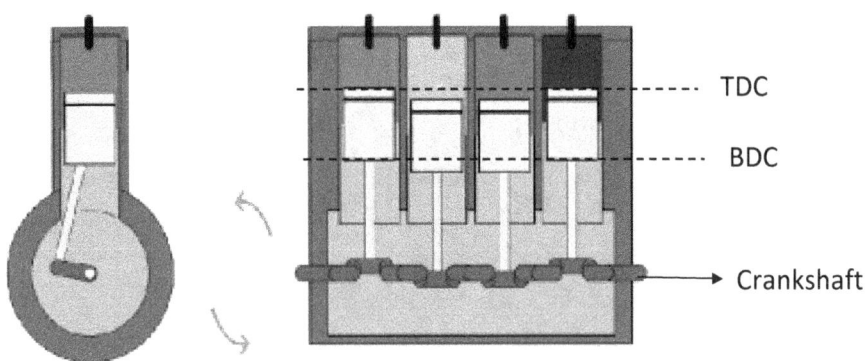

Figure 1.9.2 piston arrangement for a four cylinder engine

1st cylinder	2nd cylinder	3rd cylinder	4th cylinder	Firing order
Power	Exhaust	Compression	Induction	1
Exhaust	Induction	Power	Compression	3
Induction	Compression	Exhaust	Power	4
compression	power	induction	exhaust	2

In a four cylinder engine the firing interval will be $\dfrac{720^0 - 180^0}{4}$.

The crankshaft turns twice for a complete revolution of the camshaft which is $360^0 \times 2$ that of four cylinder engine

CHAPTER 2
Types of engine cycle

2.1 Otto Cycle (SIE)

The Otto cycle is a model of the real cycle that assumes heat addition at the top of dead center. The Otto cycle consist of four reversible cycles, which is graphically shown in figure 2.0 while the four working stroke consist of the following.

1. Induction
2. Compression
3. Power
4. Exhaust

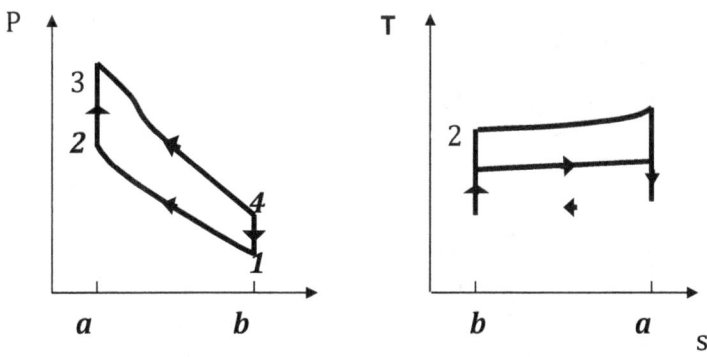

Figure 2.0 Otto cycle (p-v and t-s)

Induction stroke

The piston starts moving downward from TDC, the inlet valve opens, the displacing piston causes a partial vacuum for the intake of fresh charges (air +fuel mixture) through the inlet port into the combustion chamber, during this stroke the exhaust valve remain close.

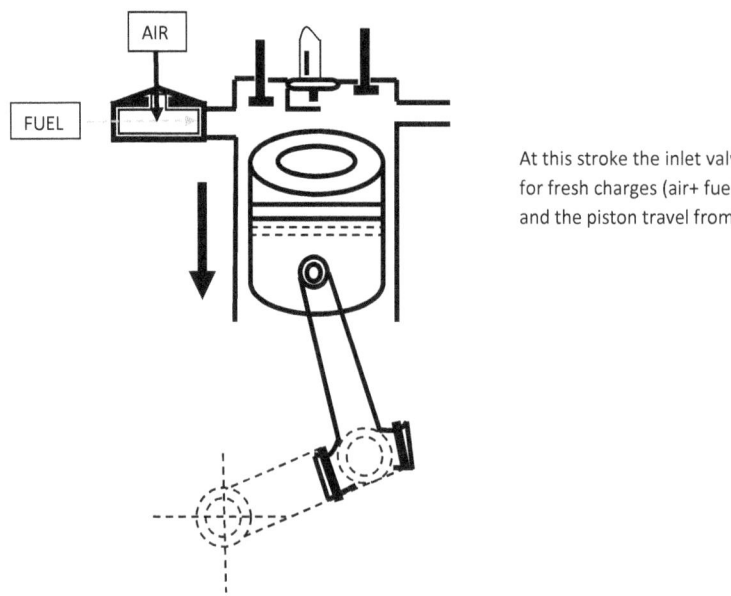

At this stroke the inlet valve will open for fresh charges (air+ fuel) to come in and the piston travel from TDC to BDC.

Figure 2.1 induction stroke

Compression stroke

At the bottom dead center (BDC) the inlet valve will close sealing the cylinder and the piston rise to compress the mixture and both valves remain close, during this stroke.

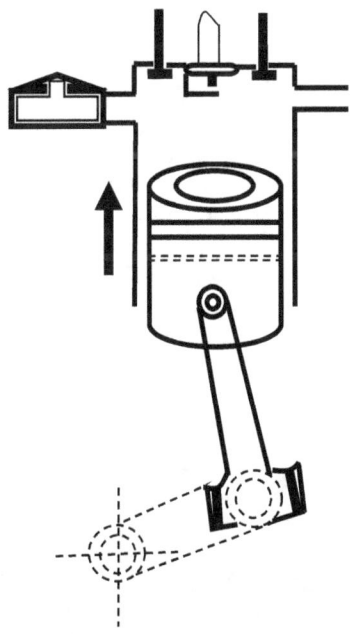

At this stroke the piston travel from BDC to TDC to compress the charges and the inlet and exhaust vale remain close.

Figure 2.2 compression stroke

Power stroke

At this stroke when the piston is close to TDC, while both valves still remain close, the compressed gas is ignited by a spark from the spark plug which will ignite the compressed gas which cause change in temperature and pressure which force the piston down to BDC.

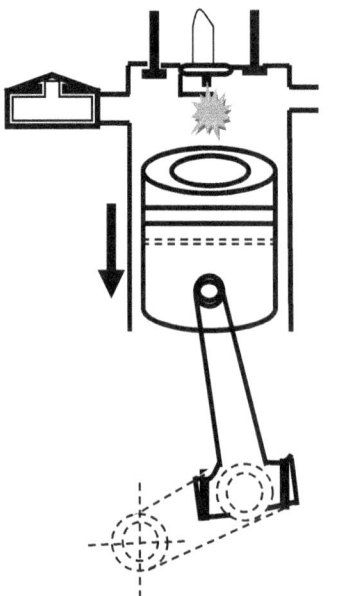

At this stroke both inlet and exhaust valves remain close the spark plug ignites a spark that forces the piston to move from TDC to BDC.

Figure 2.3 power stroke

Exhaust stroke

At bottom dead center BDC the exhaust valve open and the inlet valve closes, as the piston rises and expel out the un- burnt gas through the exhaust valve until TDC the valve closes and the piston once again commences a new induction stroke. The engine being carried over its idle strokes by the energy stored in the flywheel.

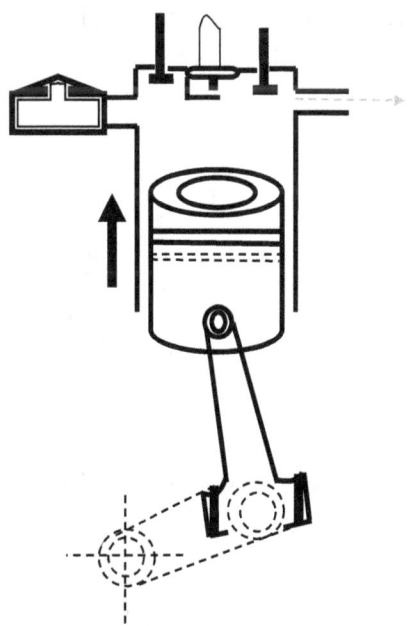

At this stroke the piston will move from BDC to TDC while the inlet vale is close and exhaust valve open to allow the un-burnt gases to go out of the combustion chamber.

Figure 2.4 exhaust stroke

2.2 Diesel Cycle

The diesel cycle is similar to Otto cycle, except that the heat addition and rejection occur at different conditions. The diesel cycle is also an ideal cycle that is it does not give an exact representation of actual process. The diesel cycle consists of four internal reversible processes. Process 1-2 is an isentropic compression. Process 2-3 is a constant pressure heat addition. This process makes the first part of the power stroke. Process 3-4 is an isentropic expansion, which makes up the rest of the power stroke. Process 4-1 finishes the cycle with a constant volume heat rejection with piston at BDC. Figure 2.5 shows the p-v and t-s diagram for the diesel cycle.

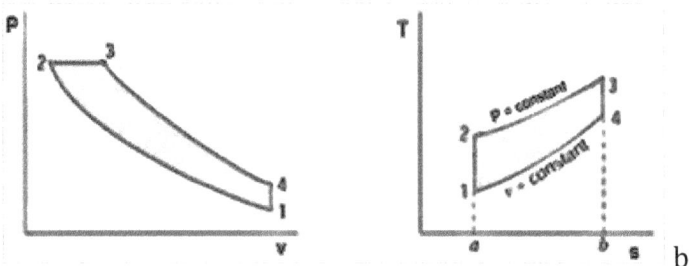

Figure 2.5 PV and TS diagram for diesel cycle

2.3 Difference between Otto and Diesel cycle

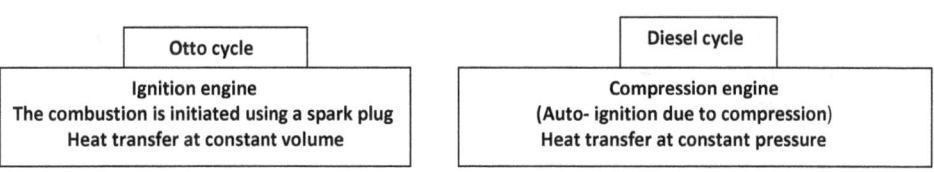

Otto cycle	Diesel cycle
Ignition engine	Compression engine
The combustion is initiated using a spark plug	(Auto- ignition due to compression)
Heat transfer at constant volume	Heat transfer at constant pressure

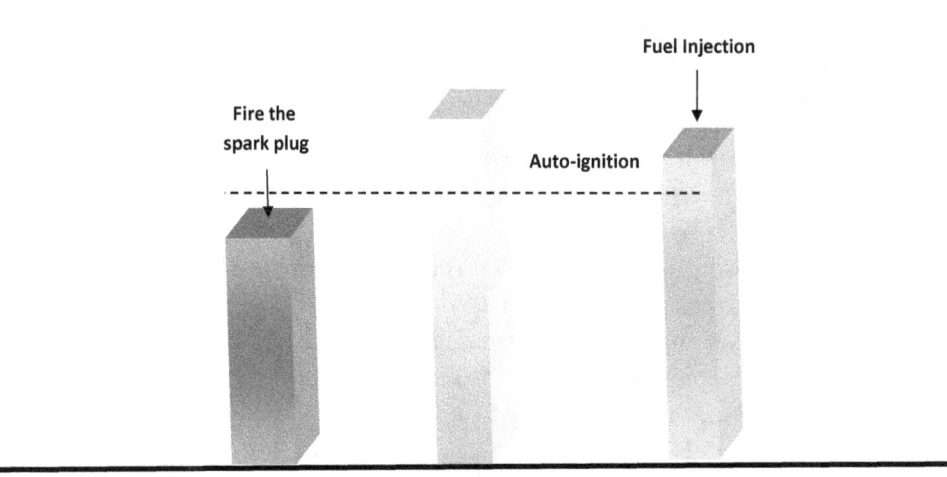

Figure 2.6 different between Otto and diesel cycle

2.4 Two Stroke Engine

The fundamental difference between the four stroke engine and the two stroke engine is the way in which the induction and exhaust process takes place.

In the four stroke engine there are separate strokes for the induction and exhaust processes. In the two stroke engine however, both the induction and exhaust processes take place with the same stroke.

The process that involves both induction and exhaust is called scavenging, or simply a gas exchange process.

The two stroke engine can be either made into a spark ignition or compression ignition engine.

The smallest engines used in two stroke engines are compression ignition engines. The engines are usually used in models and their power output does not exceed 100 W. The other type of two stroke engine with power output of up to 100 kW is spark ignition engine. Some of these engines output high power relative to their weight and bulk. Some applications of these engines are in motorcycles, chain saws and small generators.

A two stroke engine is seen in Figure 2.7. Some of the important parts of this engine are the exhaust, inlet, and crankcase port, and spark plug. The deflector is also an important design of the engine. The inlet port is where the charge is drawn from. The charge is a mixture of mainly air and fuel but may contain some exhaust. The exhaust port is where the exhaust leaves the piston, and the crankcase port provides the mixture. The combustion process for the two stroke engine goes through various processes. Following are the steps for combustion:

1) At 60^0 before hitting BDC the piston uncovers the exhaust port (EO), and the exhaust leaves the cylinder chamber while attaining atmospheric pressure. This is the end of the power stroke.

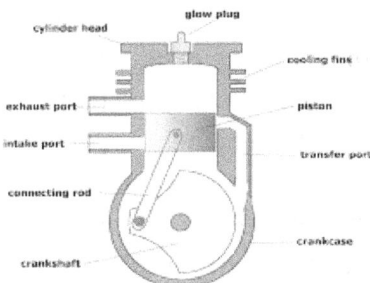

Figure 2.7 Two stroke engine.

2) At 5-10⁰ later the inlet port (IO) will open and the charge that was compressed by the crankcase will flow into the main chamber and mix with some exhaust residual. Some charge will leave the exhaust port. The deflector will aid in a way that it will divert the cross flow of charge from the inlet port into the exhaust port.

3) At about 55⁰ after BDC, with the piston moving up, the inlet port will now close (IC). There will be some back flow of charge from the inlet port into the crankcase.

4) At 60⁰ after BDC the exhaust port will close (EC) and the piston will now compress the charge through its upward movement.

5) At 60⁰ before TDC the crankcase port will open (CO) and allow charge to flow into the crankcase. The charge will flow into the crankcase since the pressure in the crankcase is below the ambient pressure.

6) When the piston is within 10-40⁰ beforeTDC the charge will be compressed enough to be at a high temperature. Then combustion will follow with flame initiation from the spark plug. In this process work is done by the engine on the air and fuel mixture. The power stroke starts when the piston hits TDC and continuous until the exhaust port opens in step.

3.1 LUBRICATION

Lubrication plays a vital role in an engine system and it also promotes the performance of engine parts. The lubricating system and the cooling system help the engine to maintain a stable operating temperature. When there is contact between two surfaces friction develops because of the relative movement between the two surfaces. Friction between metals surface causes wears. To decrease friction. The two mating surface have to be separated by a lubricant. A lubricant is a thin fluid film that separates to surface so as to reduce the friction between them.

3.2 PURPOSE OF LUBRICATION

- Reduce the frictional resistance of the engine to a minimum to ensure maximum mechanical efficiency.
- Protect the engine against wear.
- Serve as a cooling agent picking up heat.
- Remove all impurities from the lubricated region.
- Form a seal between piston rings and the cylinder walls to prevent blowby.

3.3 LUBRICATION SYSTEMS

- Mist lubrication system ⎤ Two Stroke Engines.
- Wet sump lubrication system ⎤ Four Stroke
- Dry sump lubrication system ⎦ Engines

Mist lubrication system is mainly employed in two- stroke cycle engines, whereas wet and sump systems are used in four stroke cycles engine. The wet sump system is employed in relative small engines, such as automobile engines, while the dry sump system is used in large stationary, marine and aircraft engines.

3.3.1 MIST LUBRICATION SYSTEM

in two-stroke engines, the charge is compressed in the crankcase, and as such it is not suitable to have lubricating oil in the sump, therefore such engines are lubricating by adding 3% to 6% oil in the fuel tank itself. The Oil and fuel mixture is inducted through the carburetor the fuel gets vaporized and the oil in the form of mist goes into that impinges the crankcase walls lubricates the main and connecting rod bearings and the rest of the oil lubricates the piston rings and cylinder the main advantage with this system lies in the simplicity and low cost as the system does not require any oil pump filter etc.

3.3.2 WET SUMP LUBRICATION SYSTEM

In the wet sump system the bottom of the crankcase contains an oil sump (or pan) that serves as the oil supply reservoir. Oil dripping from the cylinders and bearings flows by gravity back into the wet sump where it is picked up by a pump and re-circulated through the engine lubricating system. The types of wet sump systems used are.

- The splash and circulating pump system
- The splash and pressure system
- The full force-feed system.

3.3.3 DRY SUMP LUBRICATING SYSTEM

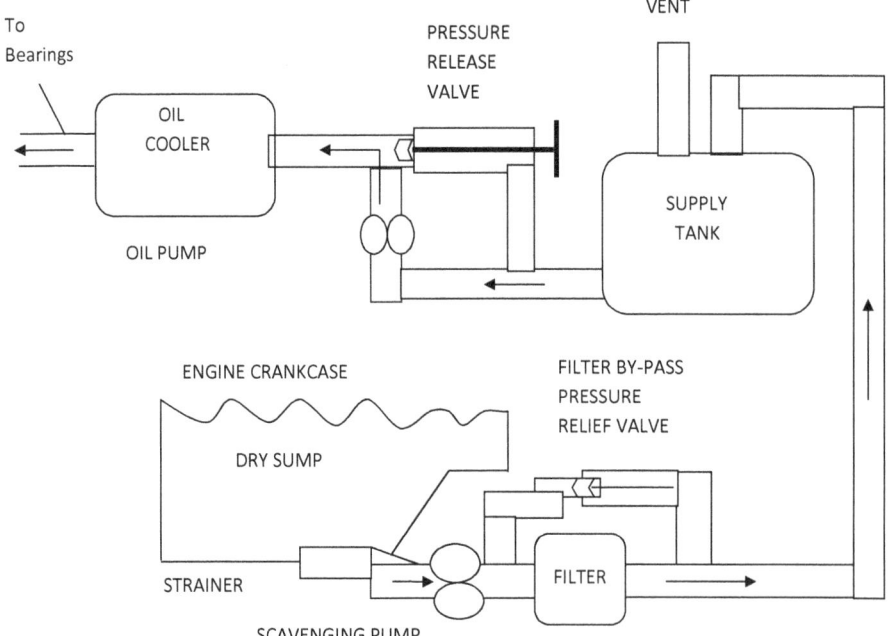

Figure 3.2 dry sump lubrication system

Dry sump lubrication system is the ultimate oiling system for internal combustion engines. The simple fact that all Formula One, Indy cars, Le Mans and Sports Racing cars as well as Super Speedway Stock Cars use dry sumps, proves this point.

The main purpose of the dry sump system is to contain all the stored oil in a separate tank, or reservoir. This reservoir is usually tall and round or narrow and specially designed with internal baffles, and an oil outlet (supply) at the very bottom for uninhibited oil supply.

3.4 PROPERTIES OF LUBRICATING OIL

- The oil used in an engine must serve as a lubricant, a coolant and an agent for removing impurities.
- It must be able to withstand high temperatures without breaking down. The oil must operate over a good range of temperature.
- The must not oxidize on the chamber walls, piston crown or at the piston rings oil should have high film strength to prevent metal-metal contact even under extreme loads.

3.5 RATING OF LUBRICATING OIL

- Lubrication oil is generally rated using a viscosity scale established by SAE.
  Commonly used viscosity grades are:
  1. SAE 5w
  2. SAE 10
  3. SAE 20
  4. SAE 30
  5. SAE 40
  6. SAE 45,
  7. SAE 50
- The oil with lower viscosity is less viscous and used in cold-weather operation. Modem high temperature high speed, close tolerance engines use high viscosity grade oil.

3.6 OIL PUMP AND TYPES

The pump's job is to draw oil from the sump and supply it with force to the moving parts of the engine. The pump must be able to force the oil into the bearing under a high pressure to keep the metal surface apart. There are three types of oil pumps commonly used in internal combustion engines.

- Gear pump
- Rotor pump
- Sliding vane pump

3.6.1 GEAR PUMP

Two gears is meshed with each other and enclosed in a aluminum pump casing. One of the gears is driven by the engine crankshaft and in turn drives the other gear.

As the gear rotate the teeth on the inlet side of the pump draw in the oil and it flows into the space left. The oil is then carried between the teeth around outer walls of the pumping chamber. When the teeth links together *again* at the outlet side of the pump the oil is squeezed out.

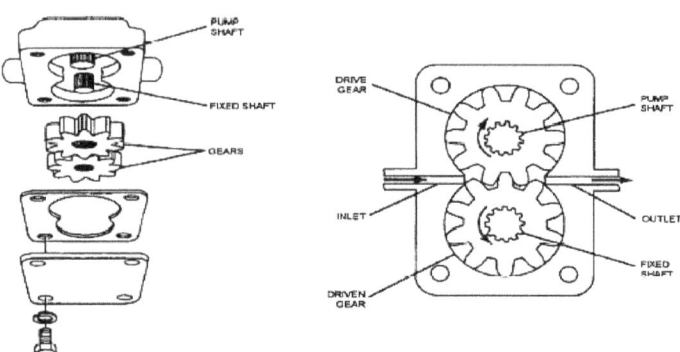

Figure 3.3   gear pump

3.6.2 ROTOR PUMP

Inside the pump there are two rotors, one inside the other. The inner rotor is mounted off-center and has one fewer lobes than the outer rotor.

The inner rotor is turned by the engine's camshaft. This turns the outer rotor at a different speed. The small space between the rotors gets larger and draws oil in through the inlet port and the large space get smaller and pump oil out through the outlet port.

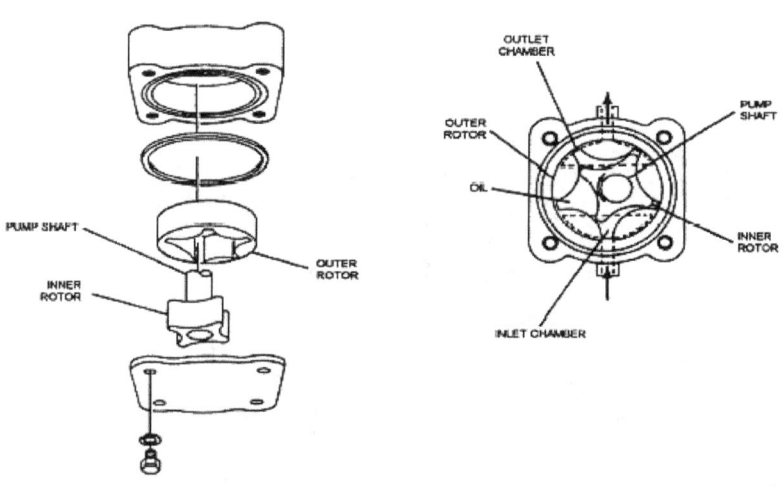

Figure 3.4 rotor oil pumps

3.6.3 SLIDING VANE PUMP

In this type of oil pimp a circular rotor is fitted off-center the pump casing. This gives a large clearance on one side, narrowing to a very small clearance on the other side. Four metal vanes are fitted in slots cuts across the surface of the rotor. These are kept in contact with the pump casing by a metal ring fitted between their inner ends.

The vane divide the chamber into four smaller chambers, which varies in size as the rotor turns, which produce the oil pumping action which is similar to the rotor oil pump.

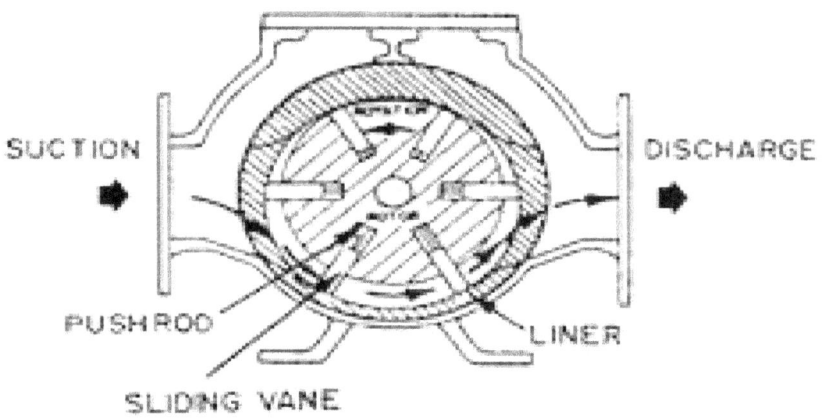

Figure 3.5 Sliding vane oil pump

4.1   COOLING SYSTEM

In internal combustion engines there are a lot of improvement in the method at which the engine is been cooled.

The major work of the cooling system it to take care of heat and also dissipate enough heat while the engine is running. It also helps to keep the engine from overheating by transferring this heat to air by a cooling fan. The car engine runs best at fairly high temperature. When the engine is cold components wear out faster and will be less efficient and emit more pollution.

There are two major type of cooling system which is named below.

- Water cooling system
- Air cooling system

4.1.1   WATER COOLING SYSTEM

The cooling system on water cooled engines circulates a fluid through pipes and passages in the engine. As the water flow through the hot parts in the engine it absorbs heat, cooling the engine. After the fluid leaves the engine it passes through a heat exchanger or radiator which transfers the heat from the fluid to the air blowing through the exchanger. In this method, cooling water jackets are provided around the cylinder, cylinder head, valve seats etc. The water when circulated through the jackets, it absorbs heat of combustion. This hot water will then be cooling in the radiator partially by a fan and partially by the flow developed by the forward motion of the vehicle. The cooled water is again recirculated through the water jackets.

4.1.2 TYPES OF WATER COOLING SYSTEM

There are two types of water cooling system which include

1. Thermo siphon system
2. Pump circulation system

Thermo siphon system

In this system water is circulated due to difference in temperature (i.e. difference in densities) of water. In this system water pump is not required but water circulates because of difference in density only.

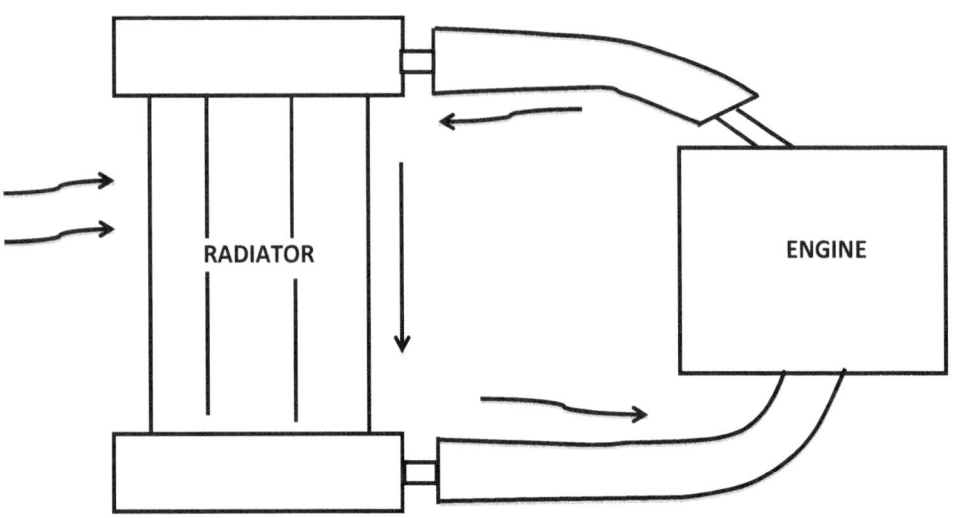

Figure 4.1 thermo siphon system of cooling

Pump Circulation System

In this system circulation of water is obtained by a pump. This pump is driven by means of engine output shaft through V-belts.

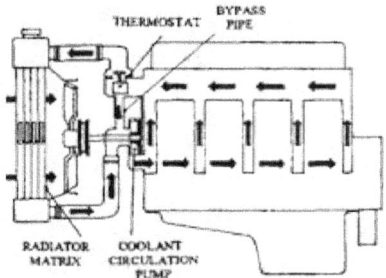

Figure 4.2 pump circulation system and component

Components of Water Cooling System

Water cooling system mainly consists of:
1. Radiator,
2. Thermostat valve,
3. Water pump,
4. Fan,
5. Water Jackets, and
6. Antifreeze mixtures.

Radiator

It mainly consists of an upper tank and lower tank and between them is a core. The upper tank is connected to the water outlets from the engines jackets by a hosepipe and the lover tank is connect to the jacket inlet through water pump by means of hose pipes.

There are 2-types of cores:

(a) Tubular core

(b) Cellular core

When the water is flowing down through the radiator core, it is cooled partially by the fan which blows air and partially by the air flow developed by the forward motion of the vehicle. As shown through water passages and air passages, water and air will be flowing for cooling purpose. It is to be noted that radiators are generally made out of copper and brass and their joints are made by soldering.

Thermostat Valve

It is a valve which prevents flow of water from the engine to radiator, so that engine readily reaches to its maximum efficient operating temperature. After attaining maximum efficient operating temperature, it automatically begins functioning. Generally, it prevents the water below 70°C.

Water Pump

It is used to pump the circulating water. Impeller type pump will be mounted at the front end. Pump consists of an impeller mounted on a shaft and enclosed in the pump casing. The pump casing has inlet and outlet openings. The pump is driven by means of engine output shaft only through belts. When it is driven water will be pumped.

Fan

It is driven by the engine output shaft through same belt that drives the pump. It is provided behind the radiator and it blows air over the radiator for cooling purpose.

Water Jackets

Cooling water jackets are provided around the cylinder, cylinder head, valve seats and any hot parts which are to be cooled. Heat generated in the engine cylinder, conducted through the cylinder walls to the jackets. The water flowing through the jackets absorbs this heat and gets hot. This hot water will then be cooled in the radiator.

Antifreeze Mixture

In western countries if the water used in the radiator freezes because of cold climates, then ice formed has more volume and produces cracks in the cylinder blocks, pipes, and radiator. So, to prevent freezing antifreeze mixtures or solutions are added in the cooling water.

The ideal antifreeze solutions should have the following properties:

1. It should dissolve in water easily.
2. It should not evaporate.
3. It should not deposit any foreign matter in cooling system.
4. It should not have any harmful effect on any part of cooling system.
5. It should be cheap and easily available.
6. It should not corrode the system.

Advantages and Disadvantages of Water Cooling System

Advantages

1. Uniform cooling of cylinder, cylinder head and valves.
2. Specific fuel consumption of engine improves by using water cooling system.
3. If we employ water cooling system, then engine need not be provided at the front end of moving vehicle.
4. Engine is less noisy as compared with air cooled engines, as it has water for damping noise.

Disadvantages

1. It depends upon the supply of water.

2. The water pump which circulates water absorbs considerable power.
3. If the water cooling system fails then it will result in severe damage of engine.
4. The water cooling system is costlier as it has more number of parts. Also it requires more maintenance and care for its parts.

4.2   AIR COOLING SYSTEM

Air cooed system is generally used in small engines say up to 15-20 kW in internal combustion engine. In this system fins or extended surface are provided on the cylinder walls, cylinder head etc. heat is generated due to combustion processes and heat will be conducted by the fins and when air flows over the fins heat will be dissipated to the air.

The amount of air dissipated to air depends on the following
1. Amount of air flowing through the fin
2. Fins surface area
3. Thermal conductivity of the metal used for the fins

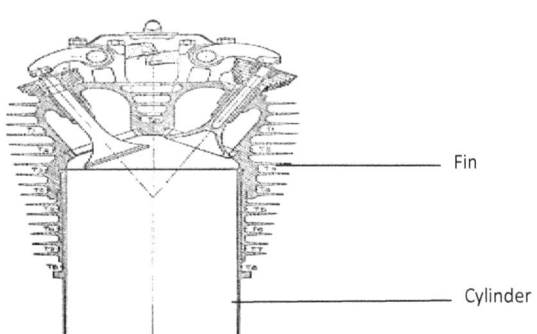

Fin

Cylinder

Figure 4.3 cylinder and fins

Advantages of Air Cooled System
1. Radiator and water pump is absent hence the system is light
2. Coolant and anti-freeze solutions are not requested
3. The system is the most suitable in cold climate, where if its water it may freeze.
4. In case of water there are leakages, but in this case there are no leakage

Disadvantages of Air Cooled System
1. Comparatively it is less efficient.
2. It is used in aero planes and motorcycle engines where the engines are exposed to air directly.

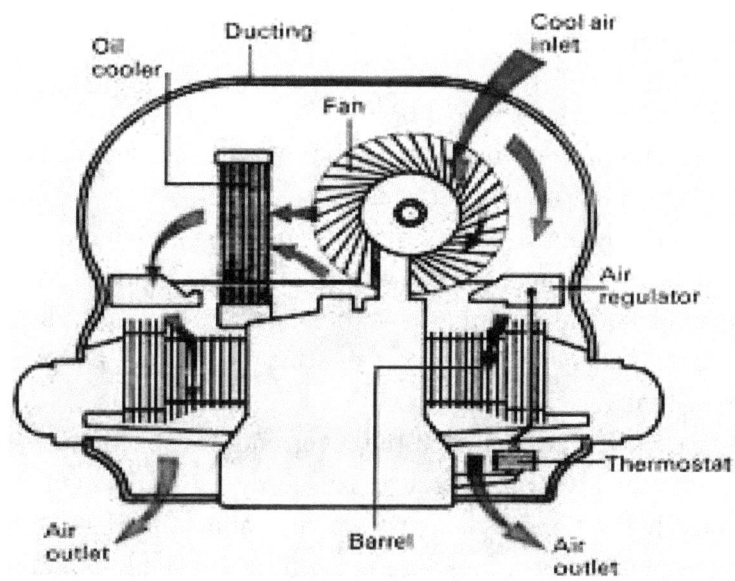

Figure 4.4 Air cooled working cycle

5.1 **FUELS SYSTEM**

In internal combustion engine the fuel (petrol and diesel) are usually stored in the tank and it convey by a pump when needed by the engine.

The tank is normally fixed far away from the engine due to the heat generated by the engine to reduce fire risk. The fuel pump helps to convey the fuel from the tank and deliver it to the unit where it will be converted to work.

It is usually fitted with a filter to stop dirt from the tank entering the carburetor or injector.

5.2   **FUEL PUMP**

There are two major types of fuel pump used in internal combustion engine.

- Mechanical pump

- Electrical pump

5.2.1 **MECHANICAL FUEL PUMP**

This type of pump is normally fitted to the side of the engine and it is operated by an extra cam via the engine camshaft.

The operating lever is pushed upward by the revolving cam. This through its pivot causes the diaphragm to be pulled down. The pressure inside the pumping chamber is now reduced and fuel is forced into it through the inlet valve by the normal air pressure inside the tank.

The cam revolves always from the lever and the spring pushes the diaphragm upward. The pressure caused by this movement closes the inlet valve and opens the outlet valve to allow the fuel to the unit needed in the engine. A fine wire mesh filter is usually fitted across the inlet pipe of the pump to stop dirt from the fuel tank to reach the rest of the fuel system where it could cause blockage.

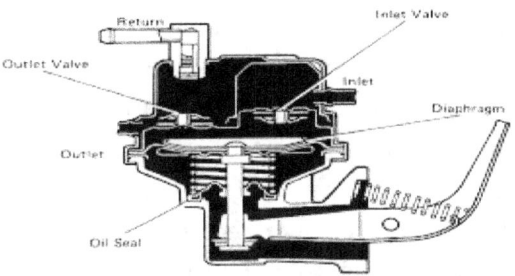

Figure 5.1 mechanical fuel pump

5.2.2 MECHANICAL PUMPS FAULTS AND THEIR CAUSES

FAULT	POSSIBLE CAUSES
Loss of pressure resulting to insufficient fuel being delivered, which may cause misfiring and loss of power at some part of the speed range and may limit engine performance.	**1.** Damage diaphragm **2.** Weak diaphragm spring **3.** Leaking valves **4.** Vapour in fuel line **5.** Broken or worn mechanical linkage
Pressure too high, causing flooding at the carburattor, increased fuel consumption and possibly stalling.	**1.** Diaphragm not properly fixed **2.** Wrong diaphragm spring

5.2.3 ELECTRICAL FUEL PUMP

The electrical fuel pump works in the same way as thed mechanical pump execp the fact that the diaphargm is moved by a solenoied instead of a cam in the engine.

When electric is swiched on, the solinoied behaves like a magnet and the amature is pilled towards it.this pull the diaphagm back and fuel is drawn into the pump through the inlet valve.

When the diaphagm is pulled back the points of the contact breaker seprates and the electric to d soid is disconnected.

A spring pushes the diaphragm back to its original position and fuel in the pump is forced out, through the outlet valve to the working unit in the engine.

The pump work as the engine is inneed of fuel in the working units.

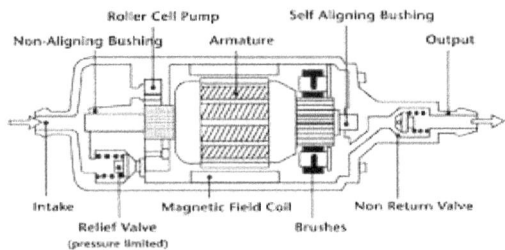

FIGURE 5.2 Electric fuel pump

5.2.4 ELECTRICAL FUEL PUMP FAULTS AND CAUSES

Pump not working	1. Check that the ignition is switch on. 2. Check circuit continuity by disconnecting the lead and check with volt meter. 3. Check that the lead to the terminal head cap is secured. 4. Examine earthing points for cleanliness 5. Examine contact point and clean, if dirty with a light wire brush. 6. Cause may be due to obstruction in suction pipe and it may be cleared by blowing air through pump intake pipe which comes from the fuel tank.
Pumps oscillates but delivers little fuel were needed	1. Choked filter or obstruction in suction side of the fuel pipe. 2. Check valve for dirt at seating flatness and wear.
pump noise	1. If the pump operates rapidly and is noisy there is probably an air leak on the suction side of the pump. Test by

	disconnecting the fuel line and immerse it in a fuel from external source and operate the pump and observe if air bubbles appear. Check all unions and joints and fuel level.

5.3 CARBURETORS

A simple carburetor only provides the correct mixture of the petrol from the tank with the right amount of air to form a spray that burns easily in the combustion chamber. A small supply of petrol from the pump is stored in the floating chamber. The position on its induction stroke draws air along the choke tube. As the air reaches the narrow part of the tube it is forced to speed up.

This increase in speed causes the pressure of the air in the venturi to fall below the pressure in the floating chamber. The difference in the two air pressure causes the petrol to flow from the floating chamber to the venturi where it is caught in the fast moving air and turned into a fine spray which is drawn into the engine. According to the direction in which the carburetor intake is located carburetors are referred to as follow.

- Downdraught

Vertical or up draught

- Side draught
- Semi-down draught

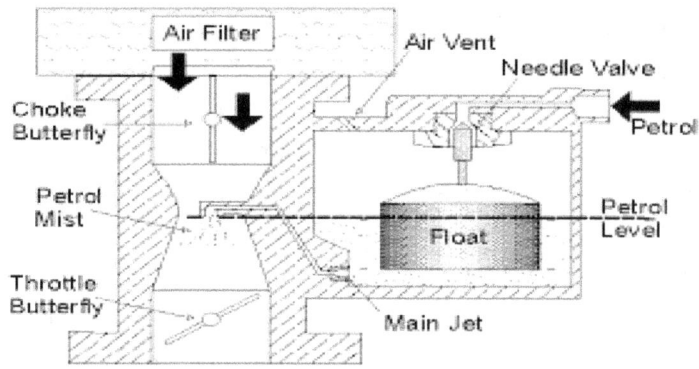

FIGURE 5.3 A carburetor

5.3.1 CARBURETOR FAULTS AND THEIR CAUSES

FAULT	CAUSES
Erratic running and stalling at idling, lucid of power high fuel consumption	Sticking piston caused by: 1. Dirty piston and suction chamber. 2. Jet out of center 3. A bent needle
Hesitation at pick-up	1. Low damper oil level requiring topping up 2. Oil with less viscosity
Float chamber flooding	1. Dirty or worn float chamber 2. Punctured float 3. Incorrect fuel level

5.4   AIR FILTER

The air drawn into the engine contains a lot of dirt and dust from the tank. This dust must not be allowed into the engine or it will cause a lot of damages to the engine. A filter is fixed to the intake of the carburetor or injector to remove the dirt. There are many types of filter, but the most common type is the paper-element, oil bath and wire-mesh.

5.5   PAPER-ELEMENT FILTER

The air is drawn through a special type of paper folded into zigzag to give a very large surface area. The dirt is left on the paper, leaving clean air to enter the engine.

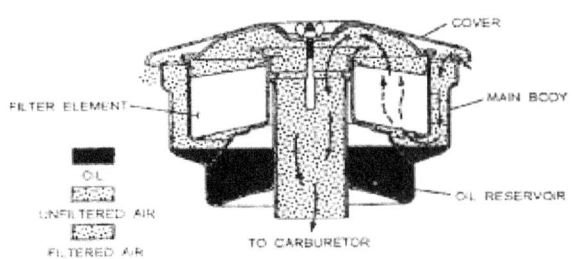

Figure 5.4 oil bath air cleaner

6.0 COMBUSTION

Combustion in compression ignition engine

In a CI engine the fuel is sprayed directly into the cylinder and the fuel-air mixture ignites spontaneously. This graph shows the fuel injection flow rate, net heat release rate and cylinder pressure for a direct injection CI engine.

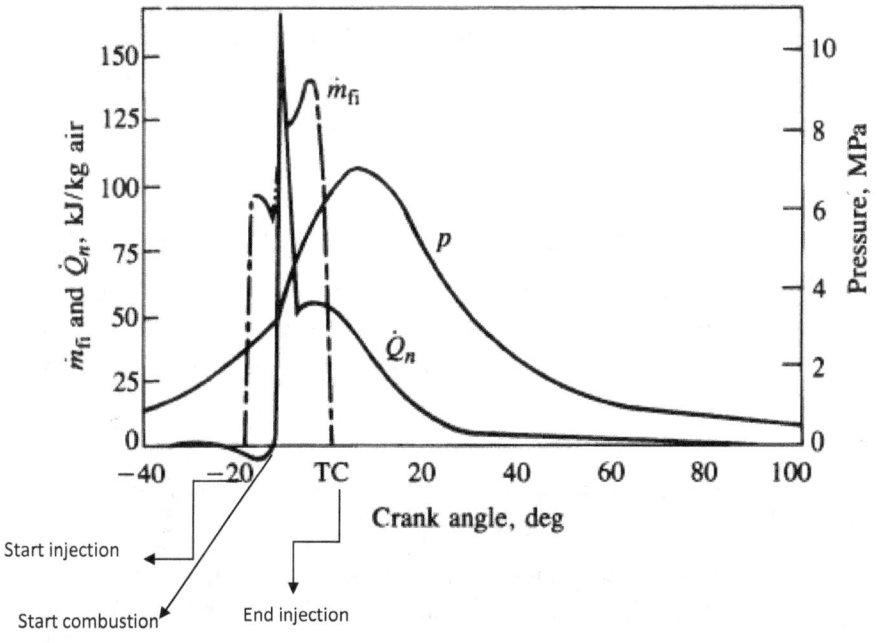

Figure 6.1 fuel injection flow rate

The combustion in ci engine is in stages which include the following:

- **Ignition delay***:* - fuel is injected directly into the cylinder towards the end of the compression stroke. The liquid fuel atomizes into small drops and penetrates into the combustion chamber. The fuel vaporizes and mixes with the high temperature high-pressure air.

- **Premixed combustion phase:** – combustion of the fuel which has mixed with the air to within the flammability limits (air at high-temperature and high-pressure) during the ignition delay period occurs rapidly in a few crank angles.

- **Mixing controlled combustion phase***:* – after premixed gas consumed, the burning rate is controlled by the rate at which mixture becomes available for burning. The rate of burning is controlled in this phase primarily by the fuel-air mixing process.

- **Late combustion phase:** – heat release may proceed at a lower rate well into the expansion stroke (no additional fuel injected during this phase). Combustion of any unburned liquid fuel and soot is responsible for this.

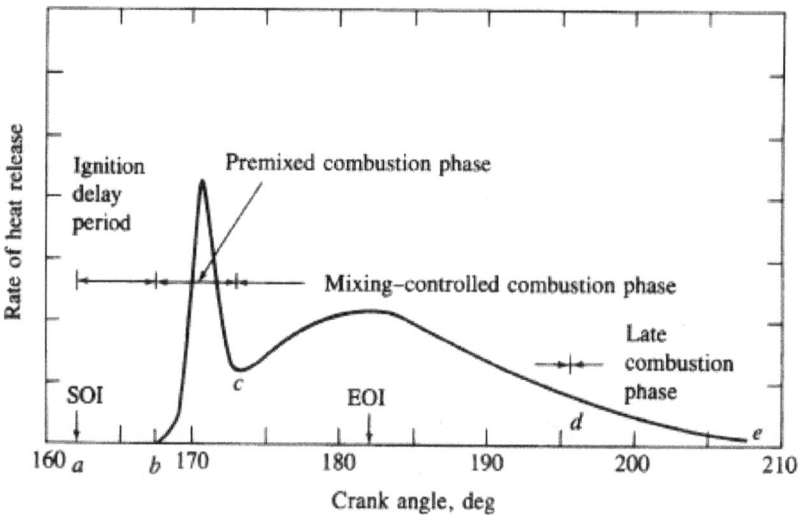

Figure 6.2 four stages of combustion in ci engine

The compression ignition engine is categorized into two types:

1. Direct injection
2. Indirect injection

1. **Direct-injection**: – have a single open head which the nozzle inject the fuel into directly into the combustion chamber. It is usually use in large stationary engine, which operates at a low speed which the time for mixing to appropriate proportion is long. So when designing a open of shallow bowl is adopted to help increase the rate of burning and engine efficiency.

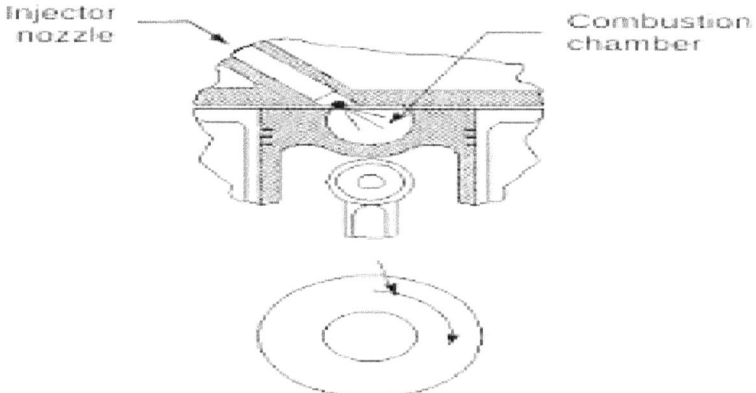

Figure 6.3 direct injection

MERIT AND DEMERIT OF DIRECT INJECTION

S/N	MERIT	DEMERIT
1	Slight lower fuel consumption due to reduced heat losses	Injection pressures are higher so wear and tear of injection equipment is higher.
2	Easier cold starting due to reduced heat losses. / No starting devices.	Injection equipment needs more frequent skilled servicing.
3	Injection pressure are relatively higher	Need good quality of fuel for good performance and clean exhaust
4	Smoother running due to highly pressure rate	To small engines, the high pressure spray is needed for atomization causes fuel to be deposited on the combustion chamber wall.

2. **Indirect-injection:** – chamber is divided into two regions and the fuel is injected into the "prechamber" which is connected to the main chamber via a nozzle, or one or more orifices. It is usually used for small high-speed engines used in automobiles chamber swirl is not sufficient, indirect injection is used where high swirl or turbulence is generated in the pre-chamber during compression and products/fuel blowdown and mix with main chamber air.

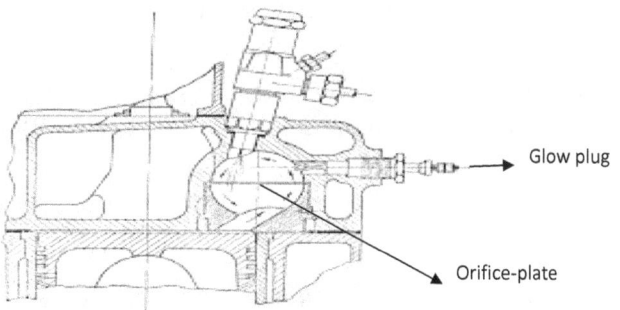

Figure 6.4 indirect injection
MERIT AND DEMERIT OF INDIRECT INJECTION

S/N	MERIT	DEMERIT
1	Less sensitive to fuel quality	Greater fuel consumption
2	Softer spray and lower wear and tear on the injection equipment	More difficult cold starting due to greater area of cold surface
3	The pintle nozzle is much less likely to become clogged or blocked than multi hole nozzle	May require cold starting aid glow plug or manifold air heater devices
4	It is suitable for small engine	Less efficient in performance

7. 0 GAS TURBINE

Gas turbine is versatile items of turbo machinery it can be used in several field of life for example power generation, aviation, marine field e.tc.

Various mechanical devices have been used to produce power for industry and society needs. Analysis on stream power plant shows that heat was added to the water and the water vapor expanded through a steam turbine, producing work. The thermal efficiency of a 500LMIN plant is about 40%. One case of the inefficiency is that an intermediate fluid, water is used to transfer the energy of the hot combustion gases to the steam turbine.

Gas turbine units overcome this by using the combustion gases directly in the turbine. A very important factor in gas-turbine selection is that gas turbine power plants are very compact and lightweight. The conventional steam power plant must occupy a far greater area and also much heavier.

Figure 7.1 a simple gas turbine

7.1 FUNDAMENTALS OF GAS TURBINE

For a gas turbine to produce any work, in hot and low pressure. Therefore, the gases must first be compressed. If after the compression the fluid is expanded through the turbine, the power produced would be used equally by the compressor, provided that both the turbine and compressor functioned ideally. If heat is added to the fluid before it reached the turbine, raising the temperature then an increase in power output should be achieved.

Unfortunately this cannot occur, the turbine blades have a metallurgical thermal limit. If the gas enters continuously higher than the temperature, the combined thermal and material stresses on the blade will cause it to inefficiency and later fail.

Typically inlet temperatures of 1300k may be found on industrial turbines.

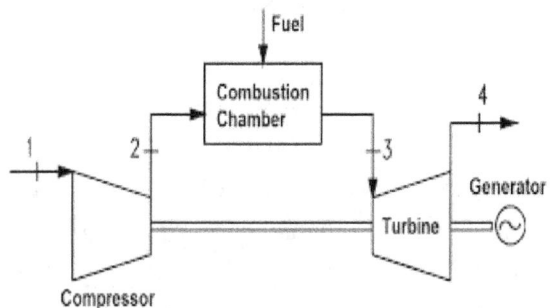

Figure 7.2 simple open cycle gas turbine

7.2   THE CYCLE ANALYSIS

The gas-turbine cycle may either be closed or open. The more common cycle is the open, in which atmospheric air is continuously drawn into the compressor, heat is added to the air by the combustion of fuel and the fluid expands through the turbine and exhausts to the atmosphere.

In the closed cycle, the heat must be added to the fluid in a nuclear power plant, and the fluid must be cooled after it leaves the turbine and before it enters the compressor.

The air-standard Brayton cycle is the ideal closed system gas-turbine cycle. It is characteristized by constant pressure heat addition and heat rejection and is entropic compression and expansion processes.

Air is the working fluid and may be considered an ideal gas. The steady-flow constant pressure processed during which heat is transferred are no longer constant temperature processes and the ideal efficiency must therefore be appreciably less than the Carnot efficiency based upon the maximum and minimum temperature of the cycle.

Also, the negative compressor work, CP $(T_1\text{-}T_2)$ is an appreciable proportion of the positive expansion work CP $(T_3\text{-}T_4)$, so that the work ratio is considerably less than Rankine cycle and it is much more susceptible to irreversibility.

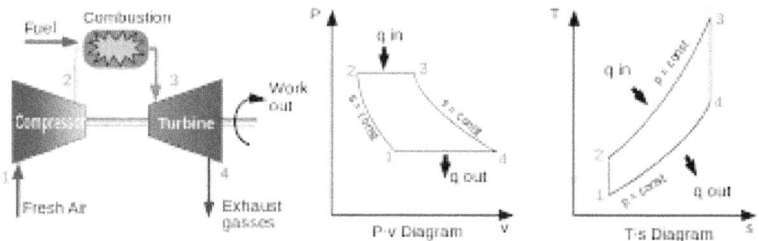

Figure 7.3 baryton cycle

The thermal efficiency 7th of the Brayton cycle may found as follows:

$7^{th} = \dfrac{Wnet}{Qin} = \dfrac{EQ}{Qin} = \dfrac{Qin-Qin}{Qin} = 1-\dfrac{Q_2}{Q_1}$

$Q_1 = mcp\ (T_3-T_2)$

$Q_2 = mcp\ (T_4-T_1)$

$7^{th} = 1-\dfrac{T_4-T_1}{T_3-T_2}$ .... equation (1)

The pressure ratio, rp is defined as:

$Rp = p_2/p_1$

And from isentropic expansion and compression processes, we find that

$\dfrac{T_2}{T_1} = \dfrac{T_3}{T_4}$

Therefore, $T_4 = \dfrac{T_3}{T_2}\ T_1$ ....equation ii

Substituting equation ii into equation i

$7^{th} = 1-\dfrac{T_3/T_2\ T_1-T_2}{T_3-T_2}$

$$7^{th} = 1 - \frac{T_1 (T_3/T_2.1)}{T_2 (T_3-1)}$$

$$7^{th} = 1 - \frac{T_1}{T_2}$$

Relating the cycle temperature to the pressure ratio

$$Tp = P_2/P_1 = P_3/P_4$$

For isentropic compression and expansion

$$\frac{T_2}{T_1} = \frac{p_2}{\{p_1\}} = rp^{\,r-1/r}$$

Or $\dfrac{T_1}{T_2} = \dfrac{1}{rp^{\,r-1/r}}$

$$7^{th} = 1 - \frac{1}{rp^{\,r-1/r}}$$

Thus, for the Brayton cycle the thermal efficiency is a function of the pressure ratio rp. The maximum temperature does have an effect on the optimum performance. If T_3 and T_1 are fixed, then there will be an optimum pressure ratio to produce a maximum amount of work, Wnet. The

Variable temperature is T_2, the temperature of the fluid leaving the compressor.

Wnet= work output from Turbine-Work input to compressor:

Work output from turbine

$(h_3-h_4) = Cp\ (T_3-T_4)$

Work input to compressor

$(h_2-h_1) = Cp\ (T_2-T_1)$

Wnet $= Cp\ (T_3-T_4) - Cp\ (T_2-T_1)$

But $T_4 = \dfrac{T_3.T_1}{T_2}$

$Wnet = Cp \dfrac{(T_3-T_3.T1-T_2+T_1)}{T_2}$

For Wnet to be maximum the $dWnet/dT_2 = 0$

$dWnet = Cp \dfrac{(T_3-T_3.T_1-T_2+T_1)dT_2}{T_2}$

$Cp \dfrac{(T_3.T_1)}{T_2{}^2} - 1 = 0$

$\dfrac{T_3.T_1}{T^2} = 1$

$T_2 = \sqrt{T_3.T_1}$

Work ratio = $\dfrac{\text{Network}}{\text{Gross work}}$

$\dfrac{Cp (T_3-T_4) - Cp (T_2-T_1)}{Cp (T_3-T_4)}$

$1-\dfrac{T_2-T_1}{T_3-T_4}$

$\dfrac{T_2}{T_1} = rp^{r-1/r} = \dfrac{T_3}{T_4}$

$T_2 = T_1{}^{rpr-1/r}$ and $T_4 = \dfrac{T_3}{rp^{r-1/r}}$

Hence, substituting

Work ratio = $\dfrac{T_1 (rp^{r-1/r}-1)}{T_3 [1-(1/rp^{r-1/r})]}$

Reference

1. Lyes kadem (thermodynamics) 2007
2. Fernando Salazar (internal combustion engine) 1998
3. Penrite oil company pty.ltd (guide to oil and greases) 2008
4. Automobile training board (inspect & service cooling system)
5. Applied thermal engineering (unit 5 cooling system)
6. Ujjwa k saha (IC lubricating system)
7. Prof. Dr. cem sorusbay(combustion in SI engine)
8. Pak piston industry @www. Pakpiston.com